Secure Web Application Deployment using OWASP Standards

An expert way of Secure Web Application Deployment

Dr.T.Subbulakshmi and Praveen Kumar H

First edition

May - 2017

Preface

This boook has been written with the objective of explaining the web security architecuture and how vulnerabilities can be analysed in the web servers.More advancced web security architecture techniques will be covered in the futrue editions. I wish all the readers for a successfull carrier in the field of web security.

Chapter one covers the introduction about the web application development and and diffrent kinds of vulnerabilities.

Chapter two discuss about the previous researchers works on the web application security and the techniques used.

Chapter three shows the syste design for the web application security, system requirements and working functionalities of the web application security.

Chapter four discuss about the implemenation and installation of systems for the Web application security.

Chapter five and six shows vulnerabilities present in the website and secure pracitces should be taken care when developing the website.

Chapter seven and eight discuss about the future directions for the work done in this research work

About the author

Dr.T.Subbulakshmi is currently working as a Professor in the School of Computing Science and Engineering at VIT University Chennai Campus, Tamilnadu, India. Earlier she has worked as the Head, Department of Computer Science and Engineering, Sethu Institute of Technology, Virudhunagar,Tamilnadu, India and Assistant Professor in Thiagarajar College of Engineering, Madurai, Tamilnadu, India.The author has contributed in framing security based curriculum and courses for engineering students like information security, network security, cryptography, intrusion detection systems and cloud security, M.Tech Computer Science and Engineering with specialization in Information Security Program. The author has completed a project Design of Masquerader Detection Systems for Information Security for Computer Society of India (CSI). The author has completed her research as a part of the Smart and Secure Environment Project, funded by National Technical Organization (NTRO),New Delhi. The author has publications in 16 Journals, 12 conferences, 3 magazines, two books to her credit.

Mr.H.PraveenKumar has completed his Master of Computer Applications in VIT University Chennai, His interest is to develop more attractive websites in all aspects. In addition to this, he is interested in analyzing vulnerabilities in the web development using Content Management Systems like Drupal and Wordpress and providing solutions to remove them. He is interested in secure web designing in compliance with OWASP standards that eradicate most of the CVE exploits. This research work done by him for showcasing vulnerabilities in the web development and the ways of fixing them would be a greater help for all the novice and expert web developers.

About the book

Web applications of todays world are facing many threats that makes the job of the security analyst a challenging one. The zero day vulnerabilities faced by the websites are one another great threat towards the protections engines. The portrait display vulnerability of software makes HP, Fujitsu and Philips notebooks is the one which was recently explored in the security world. To defend these latest and zero day attacks we need strong and round the clock mechanism that enables protection. The objective of this research is to design and develop an Application level security architecture for securing web applications against the vulnerabilities mentioned in OWASP and CVE. To illustrate the research, an event management website for Student Welfare Office of Vellore Institute of Technology Chennai Campus is developed and tested. The deployment is done using WAMP architecture, Java Script, HTML3 and CSS3 with database support enabled. This research addresses vulnerabilities mentioned in OWASP and CVE such as SQL Injection, Cross-Site Scripting, Cross-Site Request Forgery, Unvalidated Redirects and Forwards, File Upload Vulnerability and Missing Functional Level Access Control. Detection and prevention mechanism is developed for the removing the vulnerabilities and their influences in all the aspects of the web application. This website would be useful for Student welfare office of Vellore Institute of Technology, Chennai. This website will definitely be a great use for the college and the upcoming events conducted for it.

Acknowledgement

I convey my special thanks to Mr. G. Viswanathan, Founder and Chairman, VIT University, TamilNadu, India for his encouragement and support. I would like to convey my special thanks to Mr. Sankar Viswanathan, Vice President, VIT University for his encouraging words. I am thankful to Ms. Kadhambari S. Viswanathan, Assistant Vice President, VIT University for providing a diligent ambience of work.

I am thankful to Dr. Anand A Samuel for showing innovative digital pathway for achieving success in life. I am thankful to Dr. P.Gunasekaran for the sincere guidance and continued support. I am thankful to my Dean Dr. Vaidehi Vijayakumar and former Dean Dr. L. Jeganathan, Associate Dean Dr. V. Vijayakumar for providing happy work environment. I owe a deep sense of gratitude to Dr. V. Pattabiraman, Program Chair, Master of Computer Applications, VIT University Chennai, for providing necessary facilities during the course of the work. I would like to acknowledge the reviewers Dr. P. Nithyanandam, Prof.N. Ilakiya selvan, Prof. N. Hema, Prof. Umitty Srinivasa Rao and Dr. R. Rajalakshmi who are involved in the review process of the chapters for their valuable suggestions and comments. The continued support and involvement of the reviewers has helped in improving the quality and content of the individual chapters. I appreciate their work in bringing out this book.

I would like to express my thanks to Dr. A. Anantha Krishnan who is instrumental in bringing the Cognizant Technology solutions towards VIT Chennai Information Security Research Group. I wish to express my heartfelt thanks to Cognizant Technology Solutions (CTS) who has funded a laboratory for VIT Chennai named as Center for Excellence on Information Security which served as a testbed for my research towards design, development and deployment of secure web application for the college events.

Contents

List of Figures

Chapter 1

Introduction

1.1 Overview of Secure Web Application Development

A Web application widely known as Web app is an application program that is stored on a remote server and delivered over the Internet through a browser interface. Web applications use a combination of server-side scripts (PHP and ASP) to handle the storage and retrieval of the information, and client-side scripts (JavaScript and HTML) to present information to users. May it be any application for instance, we require security as a default for our application. Keeping that as the priority, the work has been taken over. The majority of web application attacks occur through cross-site scripting (XSS) and SQL injection attacks which typically comes from flawed coding, and failure to clear input to and output from the web application. Secure web application development should be enhanced by applying security checkpoints and techniques at early stages of development as well as throughout the software development lifecycle. Web applications contain resources that can be accessed by many users. These resources often traverse unprotected, open networks, such as the Internet. As most businesses rely on web sites to deliver content to their customers, interact with customers, and sell products certain technologies are often deployed to handle the different tasks of a web site.

1.2 Scope of this work

The main objective of the research is to create an event management website for Student Welfare Office for VIT (Vellore Institute of Technology) Chennai Campus. As it is for the college use it is necessary to be secure, therefore OWASP (Open Web Application Security research) is used and common vulnerabilities and exposures are encountered and fixed.

1.3 Introduction to OWASP

The Open web application security research is an open source community research set up to develop software tools and knowledge-based documentation for web application security.OWASP Top Ten is organized around particular types or categories of vulnerabilities that frequently occur in web applications. Its a list of vulnerabilities that require immediate remediation. Existing code should be checked for these vulnerabilities, as these flaws are effectively targeted by attackers. The document is not a standard or a policy. It provides a brief description of the vulnerabilities, and methods of prevention.

1.4 Top Ten Vulnerabilities

The OWASP Top 10 Vulnerabilities represents a broad consensus on the most critical web application security flaws. The errors on this list occur frequently in web applications, are often easy to find, and easy to exploit. They are dangerous because they will frequently allow attackers to completely take over user's software, steal data, or prevent user's software from working at all.

OWASP top ten vulnerabilities are,

1. Injection

2. Broken Authentication and Session Management

3. Cross Site Scripting (XSS)

4. Insecure Direct Object References

5. Security Misconfiguration

6. Sensitive Data Exposure

7. Missing Function Level Access Control

8. Cross Site Request Forgery (CSRF)

9. Using Components with Known Vulnerabilities

10. Unvalidated Redirects and Forwards

1.5 Common Vulnerability and Exposures

The Common Vulnerabilities and Exposures (CVE) system provides a reference-method for publicly known information-security vulnerabilities and exposures.CVE is a catalog of known security threats.These threats are divided into two categories:

1. Vulnerability

2. Exposures

The documentation defines CVE Identifiers (also called "CVE names", "CVE numbers", "CVE-IDs", and "CVEs") as unique, common identifiers for publicly known information-security vulnerabilities in publicly released software packages. CVE identifiers had a status of "candidate" and could then be promoted to entries, however this practice was ended some time ago and all identifiers are now assigned as CVEs. The assignment of a CVE number is not a guarantee that it will become an official CVE entry. For example: A CVE may be improperly assigned to an issue which is not a security vulnerability, or which duplicates an existing entry.

Chapter 2

Literature Review

2.1   Literature Survey

1. Title : Secure Paradigm For Web Application Development
   Author : B. Subedi, Abeer Alsadoon, P.W.C. Prasad, A. Elchouemi

   Security protection is usually thought to be a separate process in web application development phases but the external security protection mechanisms are not effective to control threats and vulnerabilities in web applications. As a consequence, researchers have realized security development should be an integral part of System Development Lifecycle of web applications. This article presents a universal secure paradigm which the web developers can apply in the development process to enhance the security features of web applications. The proposed paradigm is an extension to security development practices with agile methodology. It consists of three phases, i.e., inception, construction and transition. Inception can be mapped to analysis stage of traditional development life cycle process and transition refers to security assurance stage before deployment whereas construction phase is iterative process of security development.

2. Title : Impact of secure programming on web application vulnerabilities
   Author : Blerim Rexha, Arbnor Halili, Korab Rrmoku, Dren Imeraj

   Nowadays all organizations tend to shift their daily business processes into web. This shifting requires from web developers detailed knowledge about security techniques, such as Structured Query Language (SQL) injection and Cross Site Scripting Attack (XSS), otherwise the data managed and protected by web application could be exposed to not authorized parties. This paper aims to link and measure the impact of security techniques used by web developers for avoiding the vulnerabilities in web applications. We conducted a survey about the level of applicability of security techniques during web development and conducted a penetration testing for more than 110 local web sites. We discovered many vulnerabilities in these web sites and we linked the results with survey outcome.

3. Title : Issues, Challenges and Estimation Process for Secure Web Application Development
Author : Shivangi Gupta, Saru Dhir

The web innovation in the present scenario is experiencing an excess of uncommon changes. Today, web advancement procedure is driven by awesome proficient gatherings, yet they don't have legitimate preparation and involvement in data framework plan because of which different specialized instruments bears extreme issues. For application designer, web innovation represents a new technique of software engineering with new apparatuses, new systems and new plans. Subsequently, there's a need to locate a fitting approach to adapt up to these difficulties of web application improvement. This paper concentrates on the diverse estimation procedures and the apparatuses that are utilized for web advancement. This paper likewise connotes the different real and true issues and difficulties that ought to be taken under thought while growing expansive web applications.

4. Title : E-commerce (WEB) Application security: Defense against Reconnaissance
Author : Ashan Chulanga Perera, Krishnadeva Kesavan, Sripa Vimukthi Bannakkotuwa, Chethana Liyanapathirana, Lakmal Rupasinghe

Intrusion Detection/prevention Systems and web application firewalls provide important layer(s) of security for web applications. Even though they are well configured and maintained continually with latest attack signatures and profiles, they often fail when it comes to reconnaissance because the requests of reconnaissance to the web server often take a form of legitimate requests and they are unpredictable. Addition of signatures of reconnaissance or learning legitimate request patterns used to identify reconnaissance are practically infeasible because of the time, resource and performance issues. On the other hand IDS, IPS and WAFs prioritize attacks over the reconnaissance thus, it always tends to consider most of the reconnaissance as events not incidents which enables the adversaries to have a good understanding/profile of the web applications. The goal of this research is to analyze the reconnaissance patterns which can bypass security layers such as IDS/IPS or WAF and providing a solution which can handle the reconnaissance without hindering the performance of the application. The proposed solution is demonstrated as a plugin for a known PHP framework.

5. Title : Analysis of web application security mechanism and attack detection using vulnerability injection technique
   Author : Miss R. V. Bhor, Prof. H. K. Khanuja

The internet is growing rapidly and interconnected different wired and wireless networks with each other. By using a client server architecture computing devices which are located at different geographical locations all around the world connect to the World Wide Web. Client can access information from the web server through the web browser. Web server fetches data from the database server. Malicious minds all over the world break down the security of the data driven web applications and illegally access some private data, manipulate data or perform different malicious activities which may lead to great damage or financial loss. SQL injection attack and Denial-of-service (DOS) attack are two most important security threads found in the web applications. SQL injection is a one of the web application security vulnerability in which SQL statements are altered by attackers which is executed by the web application and submitted to the database server. DOS attack is an attack which makes network resources unavailable to its intended users. In this paper, we propose a method for evaluation of the current security mechanism by injecting vulnerabilities in the web application and exploit them using Distributed Vulnerability and Attack Detection Tool (DVADT).

Chapter 3

System Requirements

3.1 Hardware Requirements

Hardware requirement includes all the details of the hardware required for the proposed system. These are the requirements that are needed for the development of the research and the requirements that are needed for the execution of the system..

Processor	:	Intel Pentium (R), 2.16 GHz
Ram	:	2 GB
Hard Disk	:	500 GB

3.2 Software Requirements

The documents of software requirements is the system specification. It should include both a definition of a specification of requirements. It is a set of systems and what the system should do rather than the way it should. Software requirements provide a basis for the creating the software requirements specification.

Technology	:	PHP
Framework	:	Wordpress CMS
Database	:	MySQl Wamp Server
Operating System	:	Wndows 10

Chapter 4

System Design

4.1   Web Application System Architecture

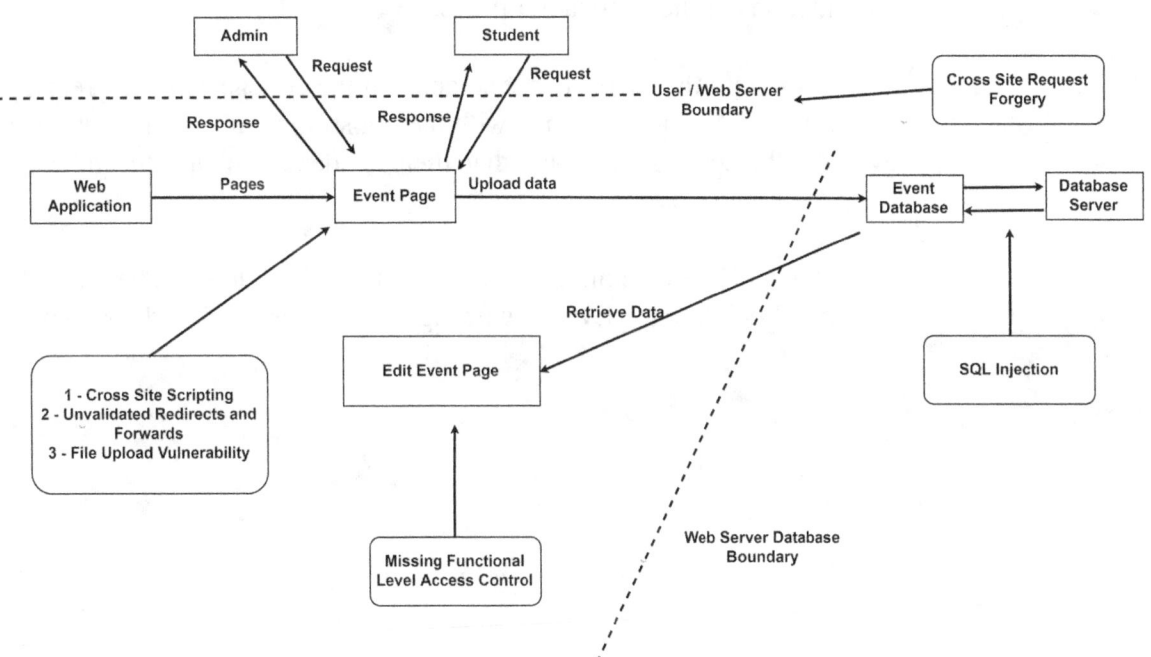

Figure 4.1: System Architecture

4.2 Content Management System

A content management system (CMS) is a software application or set of related programs that are used to create and manage digital content. CMSes are typically used for enterprise content management (ECM) and web content management (WCM). Content management is a techie task thats made easy by using WordPress built-in tools and its extensions. WordPress is a free and open-source content management system (CMS) based on PHP and MySQL.

4.3 User Role

In this research we can encounter three user roles.

Super Admin - He is the developer of the application. He owes the authority of developing, updating, managing the entire database, editing and also deleting the functions of the application if needed.

User - Here the users are the students of the college, who can register and log into the site. And few students with the registration and authorization can create events in the application. Other than them, students can view the gallery and comment their views or queries.

Admin - Here the admin is the Student Welfare Office. They have the authorization of deleting the Events, viewing the comments of all the students.

4.4   Pages

There are four different pages which need to be showcased in this research for its appropriate stages. Initially it is the basic home page which all the registered members can view. Next comes the Event page, here once you click the button it directs to another page. And finally we have the gallery and comments section.

4.5   Database

Generally database for all the applications covers the entire data. Here the data of all the students who have signed up for the application, the list of all the events and its registered student details, the data of comments from the students with their information is stored.

Chapter 5

Vulnerabilities Exist In Event Management Site

5.1 Injection

Injection is a process where any form of input is given by the attacker, that will interrupt the executing working through unintended (malicious) commands or accessing data Without proper authorization. The injection occurs when important data is sent to an interpreter (attacker) as part of a command or query. Injection may result in data loss or corruption of files, lack of accountability, or denial of access. Injection can sometimes Lead to complete host takeover.

Most useful WordPress plugins have some kind of interaction with the database. User input is frequently sent to the database, either because it needs to be stored in the DB, it needs to modify something in the DB, or because it is being used as part of a SELECT statement. If user input is not properly validated and escaped, an attacker can replace that user input with commands they can send directly to the database. There are two kinds of injection are there,

1. SQL Injection

2. Blind SQL Injection

5.1.1   SQL Injection

A classic SQL injection is a vulnerability where unfiltered user input allows the attacker send commands to the database and the output is sent back to the attacker.

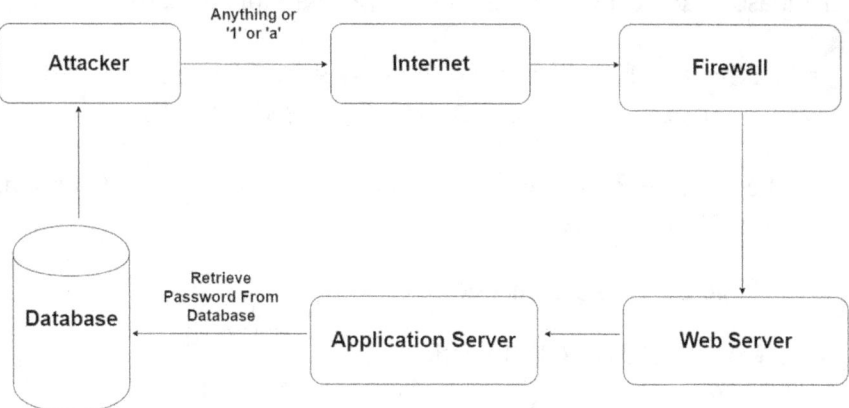

Figure 5.1: SQL Injection

Source of Attack

The following code is an example of a vulnerability caused by SQL injection. It is an SQL vulnerability because the user input in $_GET[id] is sent directly to the database without escaping. This allows an attacker to send commands directly to the database.

```
$event=$wpdb->get_results("SELECT eventname FROM".
$wpdb->eventdisplay.where ID=.$_GET['id']);
```

This type of an attack can send commands directly to the database. These include SELECT commands to download entire database including any user personally identifiable information. In some cases, it also includes INSERT and UPDATE commands to create new user accounts or modify existing user accounts.

5.1.2   Blind SQL Injection

A blind SQL injection vulnerability is when an attacker sends commands to the database but they do not get to view the database output.

For the following code a raw unsanitized user input is sent directly to the database by concatenating the $_GET[id] variable directly to the SQL query.

```
$title=$wpdb->get_var("select post_title from".
$wpdb->posts."where ID=".$_GET['id']);
```

There are generally two ways an attacker extracts data from a database using a blind SQL injection attack.

1. Time Based SQL Injection

2. Content Based SQL Injection

Time Based SQL Injection

In Time Based SQL Injection, the attacker sends the database a question like does the first letter of the first admin account start with A? If so, then sleep for 5 seconds and if not, doesnt sleep at all. If it takes less than 5 seconds for the web page to be generated and return to the web browser, then they come to an assumption that the admin account does not start with the letter A and again move on the next letter B and ask the same question. Using this technique, an attacker can send a time-based attack on a website and conclude the names of admin accounts and they can extract user passwords.

Using this technique, an attacker can send a time based attack on a website and determine the names of admin accounts and they can extract hashed user passwords.

Content Based SQL Injection

A content based blind SQL injection attack is another way for an attacker to get the data from a database when they are unable to see the database output.

For example, if the user database with ID 1 does not have a username of admin, it will however return a non-empty normal result to the browser. If the user with

ID 1 does have a username of admin. Using this technique an attacker can extract data from a database by checking for non-empty and empty responses from the application.

5.2 Cross Site Scripting

XSS flaws occur whenever an application takes un trusted data and sends it to a web browser without proper validation. XSS allows attackers to execute scripts in the victims browser which can hijack user sessions, deface web sites, insert hostile content, redirect the user to malicious sites or hijack the users browser using malware.

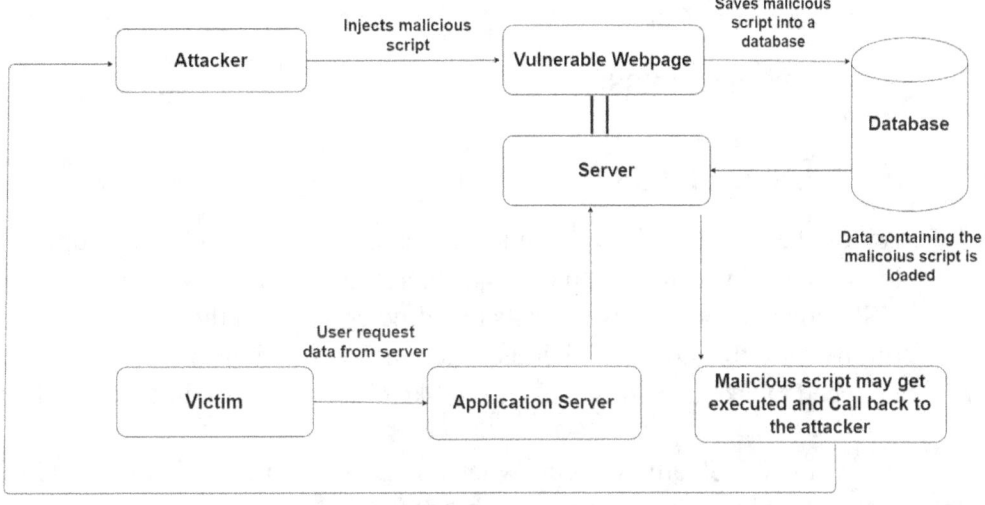

Figure 5.2: Cross Site Scripting

Source of Attack

For the following code is an example of Cross Site Scripting. It grabs a value from the URL and writes it back to the browser, unvalidated and unfiltered. If the application is hosted at https://studentwelfareactivites.tk a site visitor might visit the following URL: https:// studentwelfareactivites.tk?val=123. Visitor will see the following in the browser, Enter the value: 123 output into their browser. Probably the way the application was designed to work.

echo "Enter two values: " . $_GET['val'];

If visitor visits the following URL:https:// studentwelfareactivites.tk? val= <script>alert(XSS Script is loaded);</script>. In visitors browser the following message will be display: Enter two values: and it shows a alert box pop up saying XSS Script is loaded.

There are some three types of XSS Attack are there, namely

1. Stored XSS

2. Reflected XSS

3. DOM-Based XSS

5.2.1 Stored XSS

A stored XSS attack is an automated attack. A script can be developed that visits thousands of websites, exploits a vulnerability on each site and leaves a stored XSS payload. Anyone who visits the affected page on the site will become a victim because the stored malicious code will load in their browser. The victims do not need to take an additional action, like clicking an emailed link, to be affected.

A stored XSS attack occurs when an attacker sends malicious data to a website that is stored in a database or some other storage mechanism. Then when other site visitors visit that page or a specific URL, they are served that data which executes and performs some kind of malicious action.

5.2.2 Reflected XSS

A reflected XSS attack is usually a link that contains malicious code. When someone clicks on that link, they are taken to a vulnerable website and that malicious code is reflected back into their browser to perform some malicious action. Reflected XSS attacks are much less dangerous than stored XSS vulnerabilities.

Reflected XSS attacks rely on a victim taking some kind of action whereby they visit the target website and cause it to generate content that performs a malicious action in their browser. This makes reflected XSS attacks very difficult or sometimes impossible to automate. Each victim must be targeted individually with an email or some other content that contains a malicious link which they need to click in order to be targeted in the attack.

5.2.3   DOM-Based XSS

The attack script is based on the same page's DOM (document object model), enabling it to manipulate and interrogate it. In this type of exploit, remote execution is enabled allowing the attacker to execute malicious code on the victim's computer.

5.3   Cross-Site Request Forgery

A CSRF attack forces a logged-on victims browser to send a forged HTTP request, including the victims session cookie and any other automatically included authentication information, to a vulnerable web application. This allows the attacker to force the victims browser to generate requests the vulnerable application thinks are legitimate requests from the victim.

Attackers can trick victims into performing any state changing operation the victim is authorized to perform, e.g., updating account details, making purchases, login and logout.

Source of Attack

For the following code is an example of Cross Site Request Forgery.This code uses cookie to check if the user is logged in or not. If user is logged in then the victims session cookie and any other automatically included authentication information. This script is vulnerable to CSRF attacks. A hacker can embed the above form in any website and send the link to the users and forcing some way to click on it. If the cookie is present then the account details will be grab.

```
if(isset($_FORM["form"]) && isset($_FORM["to"])
```

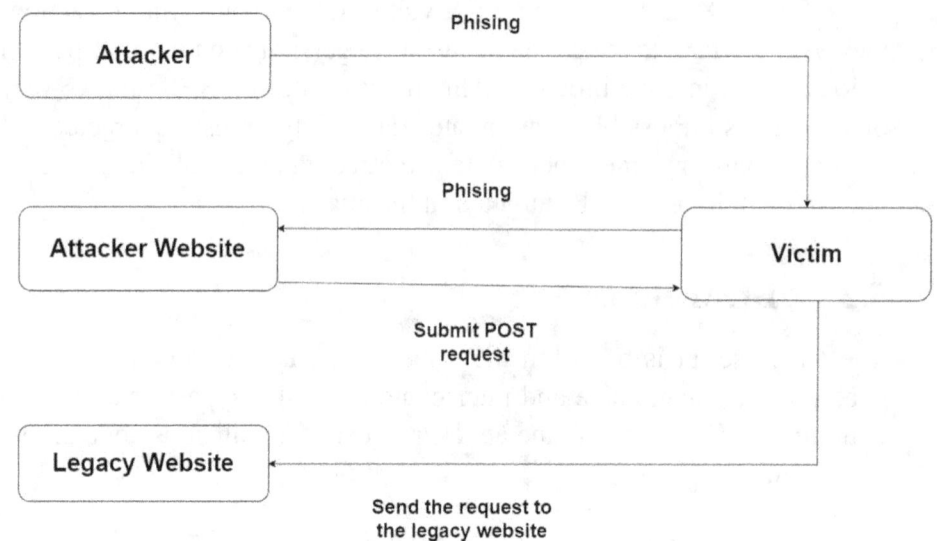

Figure 5.3: Cross-Site Request Forgery

```
&& isset($_FORM["user_details"])
&& isset($_COOKIE["user_logged_in"]))
{
    accountdetails("name", "address.", "phone");
    echo "Account Details Hacked Successfully";
}
```

5.4   Missing Functional Level Access Control

Most web applications verify function level access rights before making that functionality visible in the UI. However, applications need to perform the same access control checks on the server when each function is accessed. If requests are not verified, attackers will be able to forge requests in order to access functionality without proper authorization.

Such flaws allow attackers to access unauthorized functionality. Administrative functions are key targets for this type of attack.

Authorization means that the authenticated user has the appropriate privileges to view/control resources. Only authorized users can perform certain actions. Control access to protected resources. Prevent privileges escalation attacks.

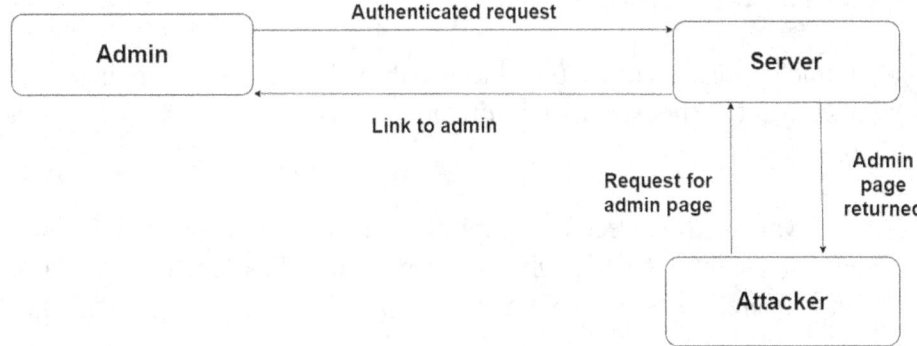

Figure 5.4: Missing Functional Level Access Control

Source of Attack

For Examle: The attacker simply forces browse to target URLs. The following URLs require authentication. Admin rights are also required for access to the https://sudentwelfareactivities.tk/login page.

If an unauthenticated user can access either page, thats a flaw. If an authenticated, non-admin, user is allowed to access the https://sudentwelfareactivities.tk/login page, this is also a flaw, and may lead the attacker to more improperly protected admin pages.

A page provides an 'action' parameter to specify the function being invoked, and different actions require different roles. If these roles arent enforced, thats a flaw.

5.5   Unvalidated Redirects and Forwards

Web applications frequently redirect and forward users to other pages and websites, and use untrusted data to determine the destination pages. Without proper validation, attackers can redirect victims to phishing or malware sites, or use forwards to access unauthorized pages.

Such redirects may attempt to install malware or trick victims into disclosing passwords or other sensitive information. Unsafe forwards may allow access control bypass.

Unvalidated redirect vulnerabilities occur when an attacker is able to redirect a user to an untrusted site when the user visits a link located on a trusted website. This vulnerability is also often called Open Redirect.

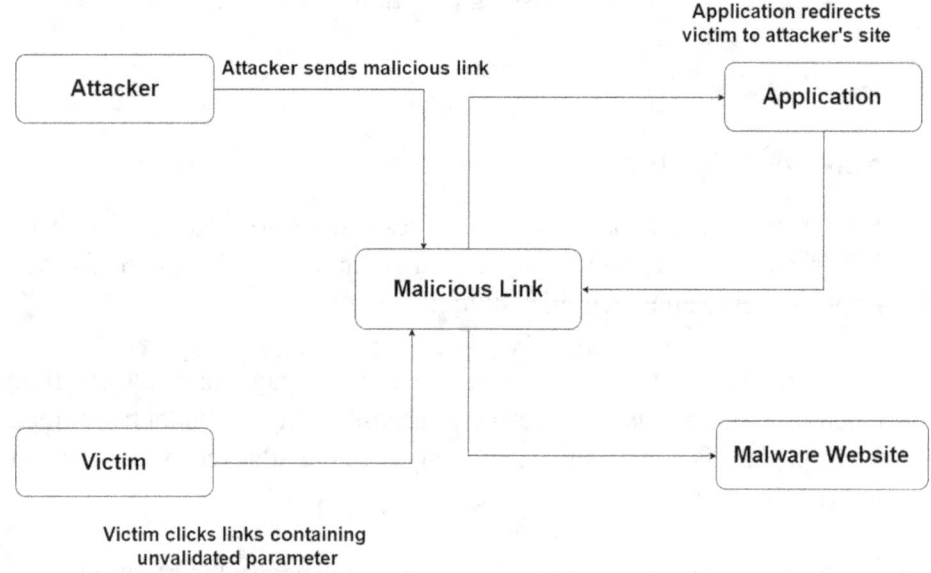

Figure 5.5: Unvalidated Redirects and Forwards

Source of Attack

All web application used to forward users to different parts of the site. In order to achieve the same, some pages use a parameter to indicate where the user should be redirected if an operation is successful. The attacker crafts an URL that will pass the application's access control check and then forwards the attacker to administrative functionality for which the attacker has not got the access. For the following code is an example of Unvalidated Redirects and Forwards.

```
<?php
header(Location: .$_GET[url);
die();
?>
```

A normal request would look something like this, https://studentwelfareactivities.tk /router.php?url=forum.php

However, as there are no checks whether the URL is internal or external an attacker would be able to conduct a URL like this as well,

https://studentwelfareactivities.tk/router.php?url=https://phishing.com

5.6 File Upload Vulnerability

Web servers follow particular criteria (file extension) to decide how to process a file. If an application allows file uploads (pictures, attached documents), ensure that the uploaded files cannot be interpreted as script files by the web server. Or the attacker may upload a script in applications programming language and execute the arbitrary code contained therein by requesting the uploaded file. Additionally, this type of attack could upload custom HTML or JavaScript files and direct a victim to them.

There are two basic kinds of file upload vulnerabilities are there,

1. local file upload vulnerability.

2. remote file upload vulnerability.

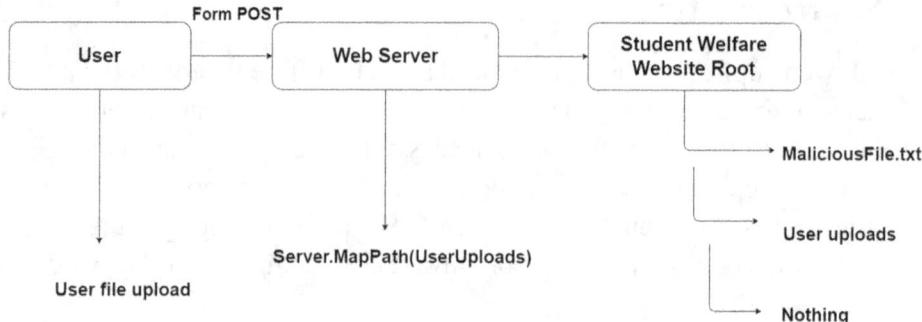

Figure 5.6: File Upload Vulnerability

5.6.1   Local File Upload Vulnerability

A local file upload vulnerability is a vulnerability where an application allows a user to upload a malicious file directly which is then executed.

Source of Attack

The following code is an example of file upload vulnerability, in this code there is no authentication or authorization check to make sure that the user has signed in and has access to perform a file upload. This allows an attacker to upload a file to the website without signing in or to have the correct permissions.

```
$file = $_FILES['wpshop_file'];
$tmp_name = $file['tmp_name'];
$name = $file["name"];
@move_uploaded_file($tmp_name, WPSHOP_UPLOAD_DIR.$name);
```

5.6.2   Remote File Upload Vulnerability

A remote file upload vulnerability is a vulnerability where an application uses user Credentials to fetch a remote file from a site on the Internet and store it locally. This file is then executed by an attacker.

Source of Attack

For an example, a remote file upload vulnerability is when an application does not accept uploads directly from site visitors. Instead, a visitor can provide a URL on the web that the application will use to fetch a file. That file will be saved to disk in a publicly accessible directory. An attacker may then access that file, execute it and gain access to the site.

In this code, there is no sanitization on the file name or contents. This allows an attacker to upload a file with a .php extension which can then be accessed by the attacker from the web and executed.

```
$fileInfo = wp_check_filetype
(basename($_FILES['wpshop_file']['name']));
if (!empty($fileInfo['ext'])) {
} else {
}
```

Chapter 6

Implementation of Secure Practices for Web Design

6.1   Injection

Prevention

Fixing these vulnerabilities of SQL Injection and Blind SQL Injection is relatively easy to use the prepare method which will automatically sanitize and escape any data that send to the database. These vulnerabilities are fixed in event display page.

This prepare() which protects SQL queries against SQL injection attacks. In short data in queries must be SQL escaped before the query is executed to prevent injection attacks. For mitigating SQL Injection is to use parameterized queries (prepare method()) or bind variables throughout the application, wherever user input is taken into consideration while forming a query. Using dynamic queries allow the user to supply input which will modify the underlying SQL query. Using parameterized queries force the application to treat all user input as data and not give any special meaning to user input. and to include blacklisting or escaping of special characters like ; - # While this can be effective in numerous cases, theres always a risk of some new attack string being discovered which will bypass these filters. The same holds true for whitelisting of certain character sets as well.

When data is included in some context, that data could be misinterpreted as a code for that environment. If that data contains malicious code, then using that data without sanitizing it, means that code will be executed. The code doesn't even necessarily have to be malicious for it to cause undesired effects. The job of sanitization is to make sure that any code in the data isn't interpreted as code.

Common vulnerabilities and exposures details,

1. CVE ID: CVE-2012-5350

2. CVE Score: 6.0

3. Vulnerability Type(s): Exec Code Sql

6.2   Cross Site Scripting

Prevention

For preventing Cross Site Scripting attacks, an application needs to ensure that all variable output in a page is encoded before being returned to the end user. Encoding variable output substitutes HTML mark-up with alternate representations called entities. The browser displays the entities but does not run them. If the input was,

```
<script>alert("Hack");</script>
```

this function would convert it into like this.

```
&quot;&gt;&lt;script&gt;prompt(Hack)&lt;/script&gt;
```

XSS Filter (Encoding Variables)

```
&           &amp;

<           &lt;

>           &gt;

/           &#x2F;
```

Data validation is the process of ensuring that a program operates on clean, correct and useful data. It uses routines, often called "validationrules" "validation constraints" or "check routines", that check for correctness, meaningfulness, and security of data that are input to the system.

Some of the validation functions are there to prevent Cross Site Scripting.

1. is_numeric(), Tests if data matches 0 to 9 with optional sign and optional decimal point

2. preg_match(), Test if data matches regular expression.

3. filter_var(), Test if data conforms to a built-in PHP filter.

4. in_array(), Tests if data is one of a range of allowed values.

In general, input validation and output sanitization and escaping will make application safe. These vulnerabilities are fixed in Create Event and Contact Form Page.

Common vulnerabilities and exposures details,

1. CVE ID: CVE-2014-5240

2. CVE Score: 2.1

3. Vulnerability Type(s): XSS

6.3   Cross-Site Request Forgery

Prevention

For preventing Cross Site Request Forgery from three ways,

Checking for Referral Header

Checking for a referral header can help in preventing the CSRF. If the request is coming from some other domain, it must be the fake request so block it. Always allow requests coming from the same domain. This method fails if the website has open redirection vulnerabilities. Attackers can perform GET CSRF by using open redirection. Most of the applications use HTTPS connection. In this the referrer will be omitted. So this method will not help if a website is using https.

Captcha Verification in Forms

Captcha verification process was initially developed to prevent BOT spam in forms. But it can also be helpful in preventing CSRF. As the captcha is generated on the client side randomly, an attacker cannot guess the pattern. So, attacker will never be able to send the correct Captcha with a fake request and all fake requests will be blocked by a Captcha verification function.

Token Pattern

This is the most secure method for preventing CSRF. In this method, the website generates a random token in each form as a hidden value. This token is associated with the users current session. Once the form is submitted, website verifies whether the random token comes via request. If yes, then verify whether it is right. By using this method easily identify whether the request was made by the user of attacker.

This vulnerability which is fixed in login page. Common vulnerabilities and exposures details,

1. CVE ID: CVE-2017-6819

2. CVE Score: 4.3

3. Vulnerability Type(s): CSRF

6.4   Missing Functional Level Access Control

Prevention

For preventing Missing Functional Level Access Control, if an application has failed to properly restrict function level access is to verify every application function.Using a proxy, browse application with a privileged role. Then revisit restricted pages using a less privileged role. If the server responses are alike, probably vulnerable.

Access control highly recommended to always apply a deny-by-default rule. Disallow access to all functions in the app by default, then allow access only to those users and other parts of the application that need it. Even when access to functions within a web application are allowed, each request needs to be verified at the time of the access. Check that the requests are from valid authorised users.

Common vulnerabilities and exposures details,

1. CVE ID: CVE-2017-7589

2. CVE Score: 4.0

3. Vulnerability Type(s): +Info

6.5   Unvalidated Redirects and Forwards

Prevention

In order to avoid malicious (unvalidated redirects and forwards), a web application can make various checks. It could allow only certain URLs/pages to trigger the redirect, or it could check if the URL matches a certain pattern (this is typically done through regular expressions). There cannot be a single valid solution to preventing unvalidated redirects, as what needs to be done depends on the needs of the web application and the redirecting endpoint but measures must be taken to disallow potential attackers from redirecting the web applications users to arbitrary web sites.

Avoiding the use of redirects and forwards. Not to involve user parameters in calculating the destination. If destination parameters cant be avoided, ensure that the supplied value is valid, and authorized for the user. It is recommended that any such destination parameters be a mapping value, rather than the actual URL or portion of the URL, and that server side code translate this mapping to the target URL.

Common vulnerabilities and exposures details,

1. CVE ID: CVE-2016-9028

2. CVE Score: 5.8

3. Vulnerability Type(s): Unvalidated Redirects and Forwards

6.6   File Upload Vulnerability

Prevention

Content-type Verification

Securing file upload forms to check the MIME(Multipurpose Internet Mail Extensions) type any uploaded file that returns from PHP. The $_FILES[uploadedfile] [type] holds the value of the MIME type, and is equal to specific values and which permits users to upload. If it is equal, the file will be uploaded, if not, the file will not be uploaded and it shows error message.

File Name Extension Verification

Securing file upload forms that uses a blacklist approach to create a list of dangerous extensions, and access will be denied if the extension of the file being uploaded is on the list.

Image File Content Verification

Securing file upload typically check if the function returns a true or false and validates any uploaded file using this information. If a malicious user tries to upload a simple PHP shell embedded in a jpg file, the function will return false. getimagesize() function to validate the image header. The getimagesize() function takes a file name as an argument and returns the size of the image if its true, and relays the indication false if the argument has failed.

File upload vulnerability preventions,

1. Allows only specific file extensions.

2. Allows only authorized and authenticated users to use the feature.

3. Checks any file fetched from the Web for content.

4. Serves fetched files from application rather than directly via the web server.

5. Stores files in a non-public accessibly directory.

6. Writes to the file when data store it to include a header that makes it non-executable.

Common vulnerabilities and exposures details,

1. CVE ID: CVE-2015-4133

2. CVE Score: 7.5

3. Vulnerability Type(s): File Upload Vulnerability

Chapter 7

Results and Discussion

SQL Injection Attack

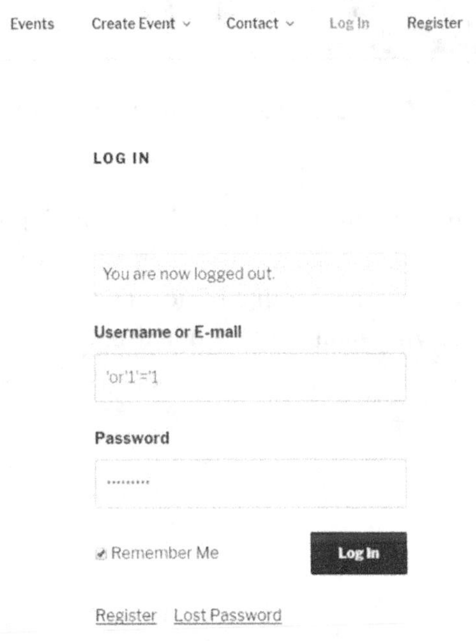

Figure 7.1: Execution of SQL Injection

SQL Injection Executed

Figure 7.2: SQL Injection Executed

When the malicious user tries to inject a malicious code to access database without their victims knowledge. The command which is sent successfully executed and the malicious user logged into the admin login, that malicious can do anything like modifying data, delete database.

SQL Injection Prevention

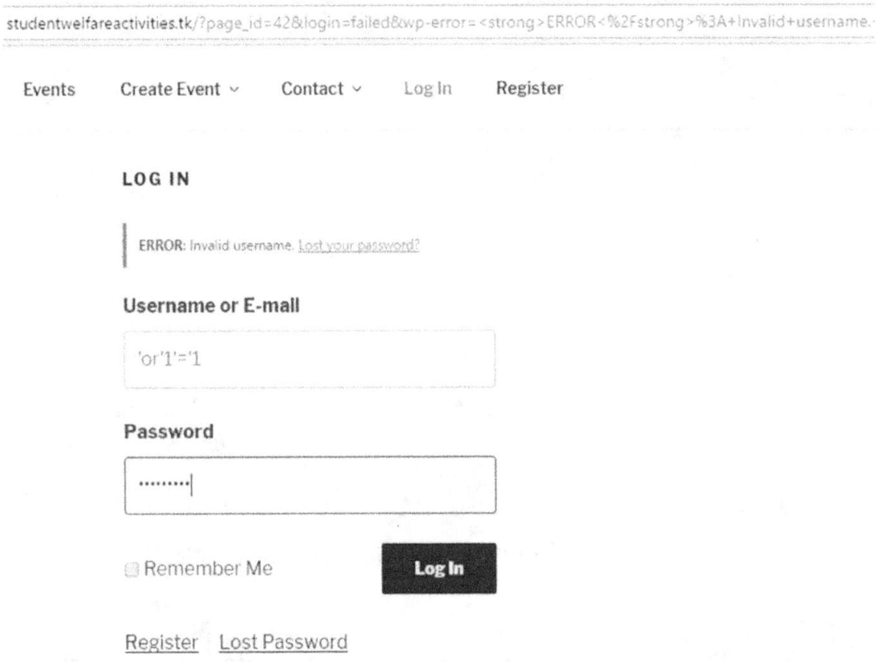

Figure 7.3: SQL Injection Prevention

When the malicious user tries to execute a SQL injection query, the prepare() method which deny the query and it show error message as a login failed. This method which will automatically sanitize and escape any data that send to the database.

Cross Site Scripting Attack

Figure 7.4: XSS Script Executed

When the malicious user tries to send a malicious script in unvalidated form, that malicious script which is executed successfully and the database contain malicious script is loaded. In that event page the alert pop up message displays as a Cross Site Scripting.

XSS Prevention

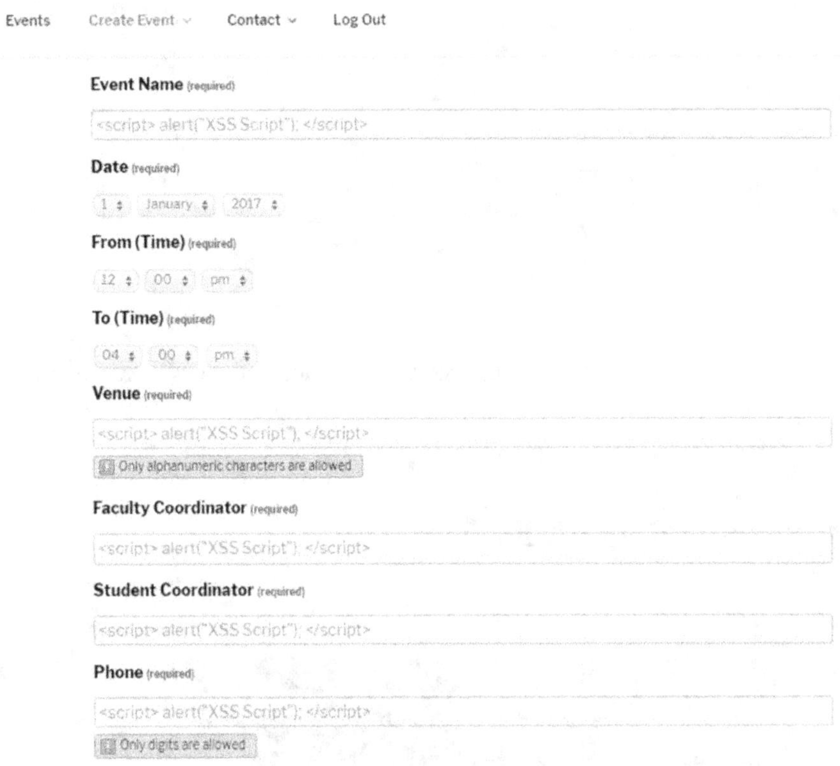

Figure 7.5: XSS Prevention

When the malicious user tries to execute a XSS script, the encoded variables which deny the script and it shows error message as, only digits are allowed, only alphanumeric characters are allowed.

Cross Site Request Forgery Attack

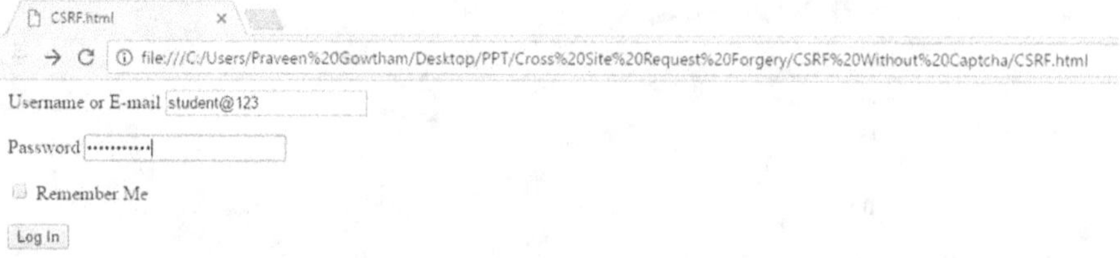

Username or E-mail student@123

Password ••••••••••

☐ Remember Me

Log In

Figure 7.6: Cross Site Request Forgery Execution

Cross Site Request Forgery

Figure 7.7: Cross Site Request Forgery Executed

When the malicious user tries to send a forged HTTP request, that request which successfully accepted and logged in, that malicious user can perform any state of changing details updating event, delete event and edit event.

Cross Site Request Forgery Prevention

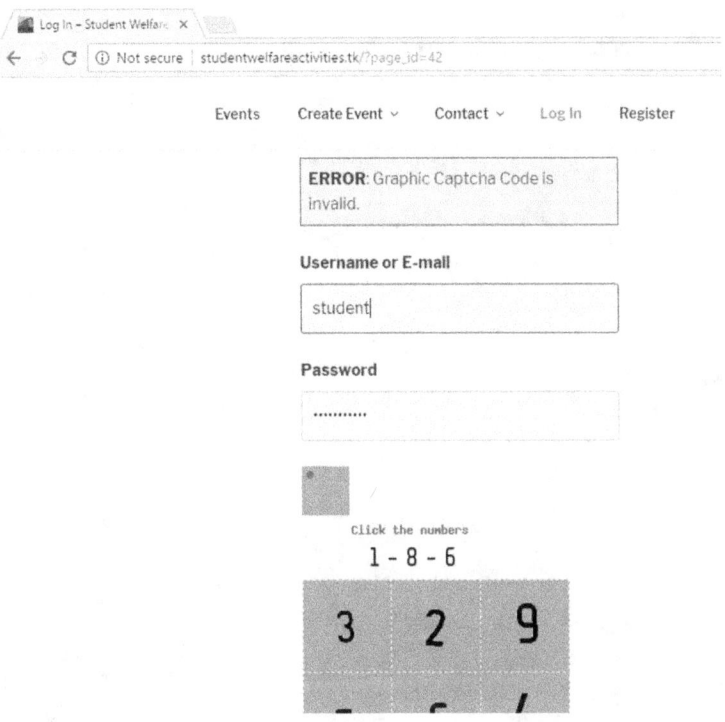

Figure 7.8: Cross Site Request Forgery Prevention

When the malicious user tries to attack a logged-on browser to send a forged HTTP request, but an captcha which deny the forged HTTP request.

Missing Functional Level Access Control

Events   Create Event ˅   Contact ˅   Log In   Register

LOG IN

You are now logged out.

Username or E-mail

student

Password

••••••••••

☐ Remember Me   **Log In**

Register | Lost Password

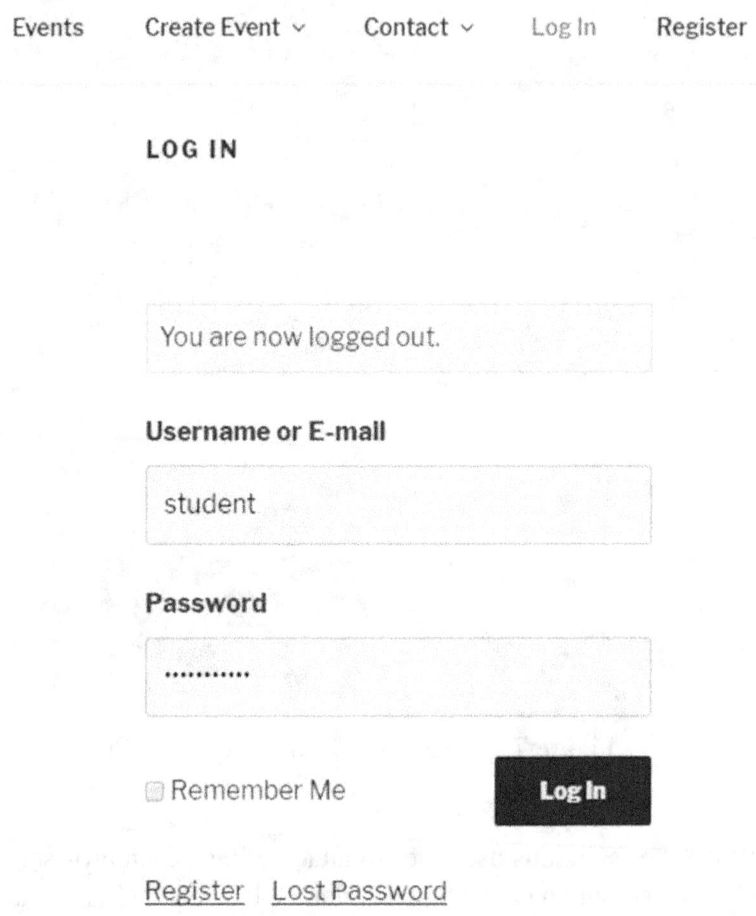

Figure 7.9: Execution of Missing Functional Level Access Control Executed

Deleting Events

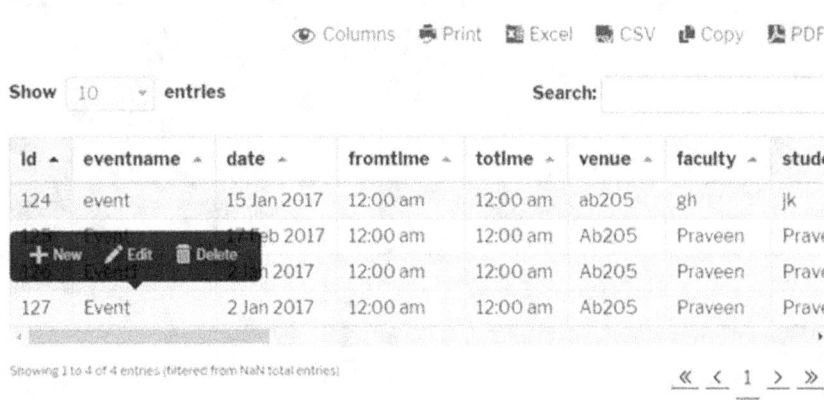

Figure 7.10: Missing Functional Level Access Control Executed

When an unauthorized person tries to login the edit event page, that page which is not properly verify the privileges and unauthorized person logged into that page and deleting events.

Missing Functional Level Access Control Prevention

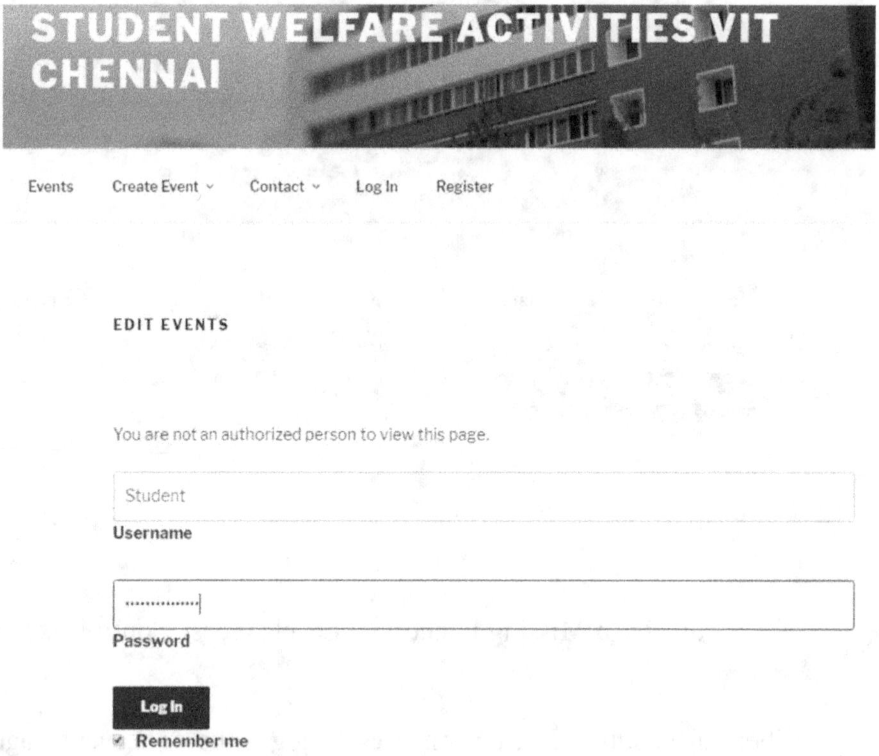

Figure 7.11: Missing Functional Level Access Control Prevention

When an unauthorized person tries to authenticate the page, the functional level which verify the privileges and deny the unauthorized access.

Unvalidated Redirection and Forwards

Figure 7.12: Execution of Page Redirection

Page Redirected

Figure 7.13: Page Redirected

When the malicious user tries to execute a malicious script to redirect page. The script which is executed and redirected to another page. Without proper validation, the website which takes untrusted data and redirected to another page.

Unvalidated Redirects and Forwards Prevention

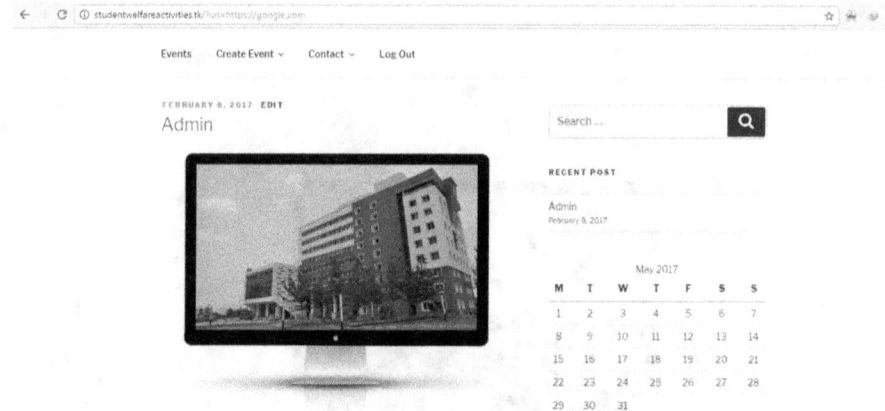

Figure 7.14: Unvalidated Redirects and Forwards Prevention

When the malicious user tries to redirect a page, the server side code which translate mapping value to the target URL and which deny the page redirection and that page which remains in same state.

Quform Multiple Parameter Forbidden Error

CREATE EVENT
Edit this form with Visual Composer

Event Name (required)

Date (required)
Day ▾   Month ▾   Year ▾

From (Time) (required)
12 ▾   00 ▾   am ▾

To (Time) (required)
12 ▾   00 ▾   am ▾

Venue (required)

Faculty Coordinator (required)

Student Coordinator (required)

Phone (required)

Upload Poster (required)
Browse
Maximum size 25MB

Upload Event Info (required)
Browse
Maximum size 25MB

Submit
Edit this form

Figure 7.15: Quform

This is an Create Event Page, The user can upload an event details like Event Name, Date , Time, Faculty Coordinator, Student Coordinator, Phone, Event Info, and Event Poster.

Event Display

| 125 | Event | 17 Feb 2017 | 12:00 am | 12:00 am | Ab205 | Praveen | Praveen | 8870103095 | | |

Figure 7.16: Event Info Error

Multiple Parameter Forbidden Error

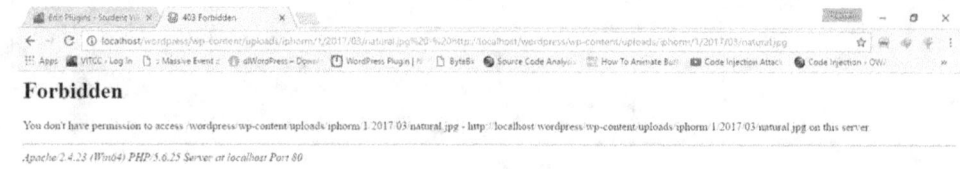

Figure 7.17: Multiple Parameter Forbidden Error

When the user tries to upload an event details, the form which sumits the details and it will redirect to another page in event display page. In that page the image doesn't shown. In that particular page the multiple parameter error which occurs and it shows as error message like Forbidden.

Event Display

Figure 7.18: Image Shown

After Prevention of Multiple Parameter Forbidden Error

Figure 7.19: After Prevention of Multiple Parameter Error

When the user tries to upload an event details, the form which sumits the details and it will redirect to another page in event display page. In that page the image is shown. In that particular page the multiple parameter error which is fixed.

File Upload Vulnerability Attack

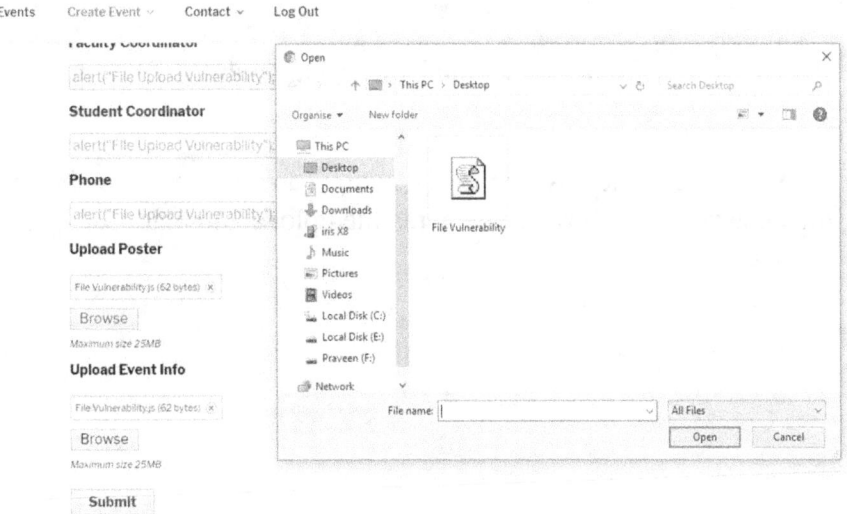

Figure 7.20: Malicious File Executed

When the malicious user tries to execute a malicious script to upload a malicious file, the unvalidated form which takes malicious file and successfully uploaded. This may occurred in database contain malicious file.

File Upload Vulnerability Prevention

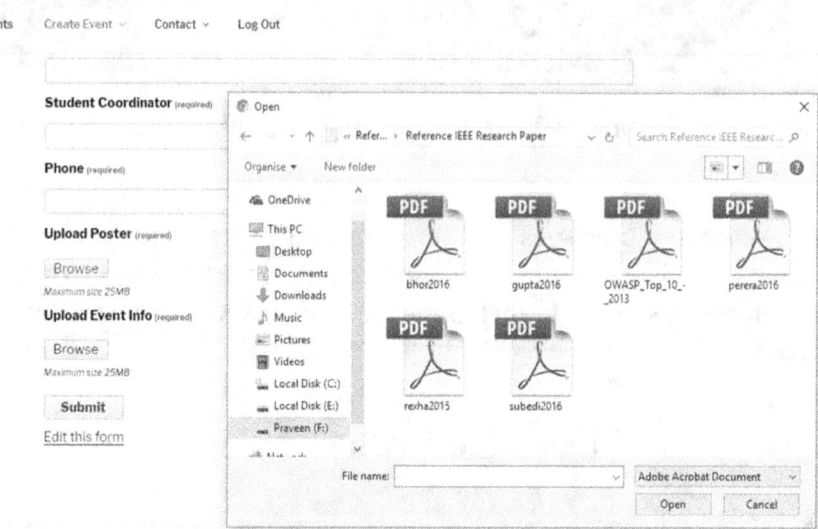

Figure 7.21: File Upload Vulnerability Prevention

When the malicious user tries to upload an malicious file, the content type and filename verification which deny the file upload vulnerability.

Chapter 8

Conclusion and Future work

The research has been taken over with the concern of the security as the priority. Hence ,OWASP one of the top securities has been used for the application. Therefore the web application is quite secure for the management to use and functionalize. This application is user friendly and is ready to use. If needed for further updates the site can be reprocessed by the Super Admin. Future updates, file encryption, page encryption and password encryption.

Appendices

Appendices A

Various Snapshots

Login Page

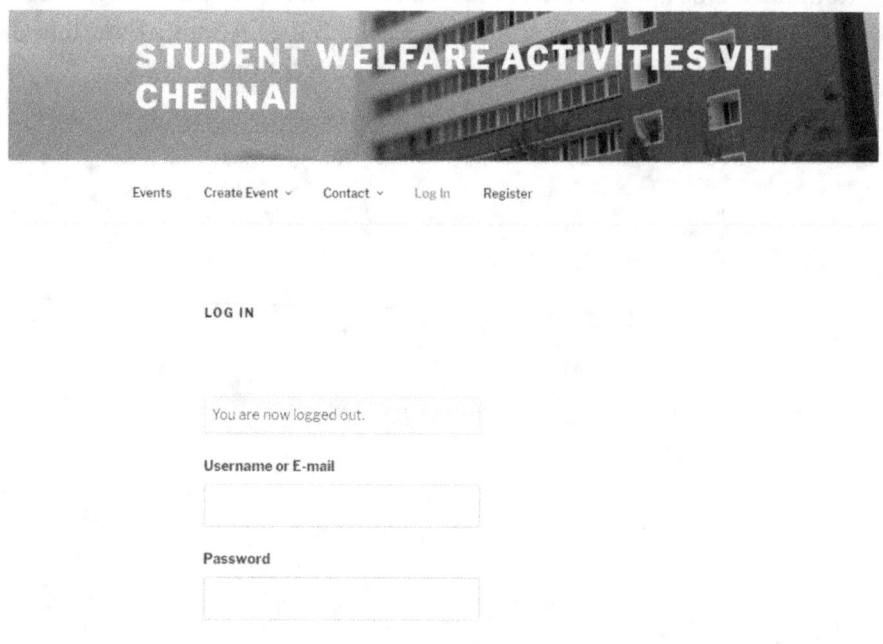

Figure 8.1: Login Page

Registration Page

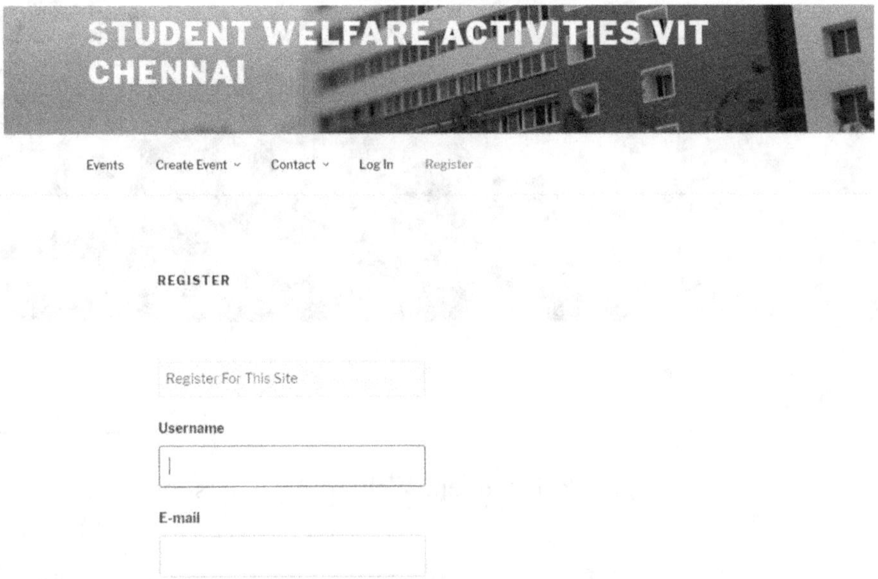

Figure 8.2: Registration Page

Student Welfare Activities Index Page

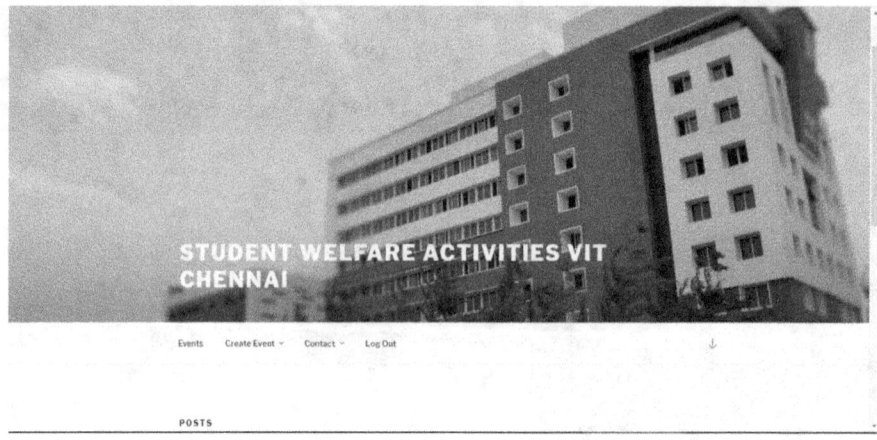

Figure 8.3: Student Welfare Activities Index Page

Event Display Page

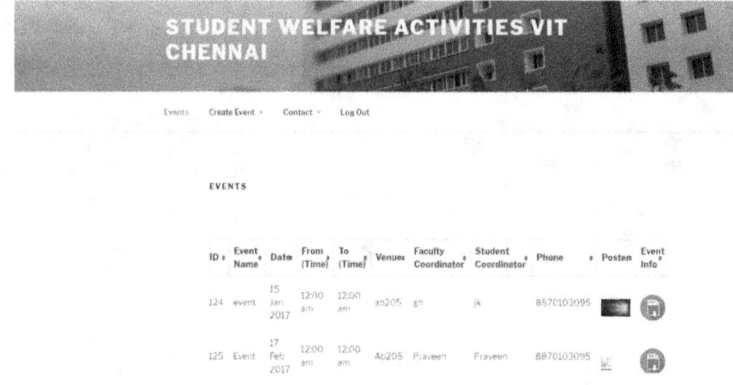

Figure 8.4: Event Display Page

Create Event Page

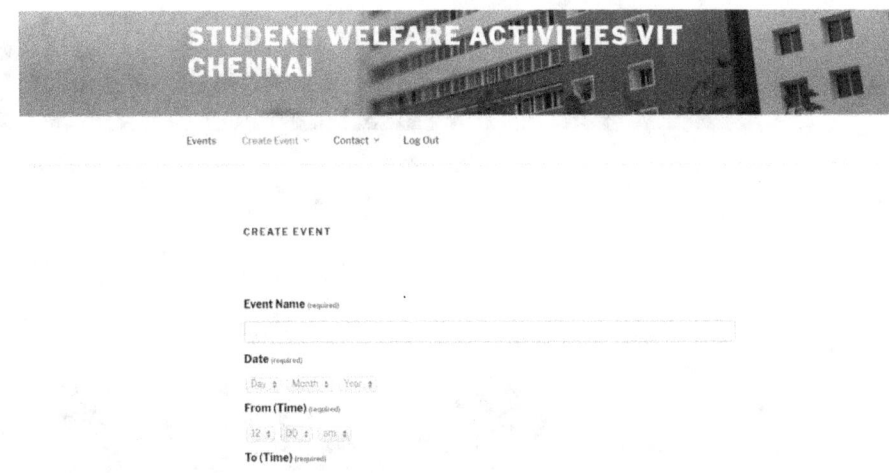

Figure 8.5: Create Event Page

Edit Event Page

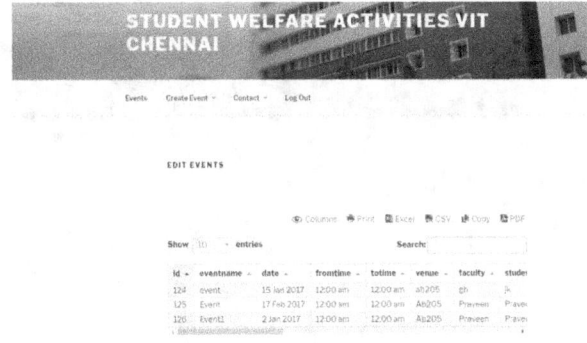

Figure 8.6: Edit Event Page

Events Gallery Page

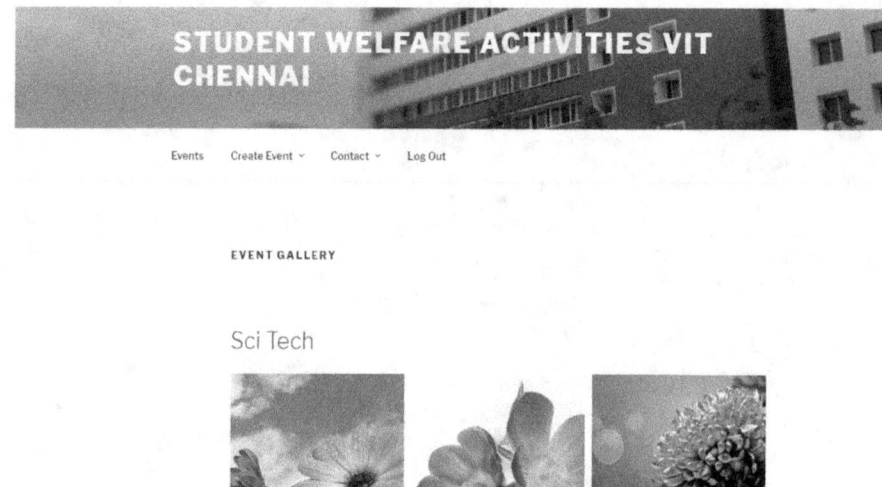

Figure 8.7: Events Gallery Page

Contact Form Page

Figure 8.8: Contact Form Page

Comments Page

Figure 8.9: Comments Page

Bibliography

References

[1] Ashan Chulanga Perera, Krishnadeva Kesavan, Sripa Vimukthi Bannakko-tuwa, Chethana Liyanapathirana, Lakmal Rupasinghe, E-commerce (WEB) Application security: Defense against Reconnaissance, Department of Information Systems Engineering, Sri Lanka Institute of Information Technology, Malabe, Sri Lanka.

[2] B. Subedi, Abeer Alsadoon, P.W.C. Prasad, A. Elchouemi, Secure Paradigm For Web Application Development, School of Computing and Mathematics, Charles Sturt University, Sydney, Australia 2Walden University, 2016 IEEE Paper.

[3] Blerim Rexha, Arbnor Halili, Korab Rrmoku, Dren Imeraj, Impact of secure programming on web application vulnerabilities, Faculty of Electrical and Computer Engineering University of Prishtina Prishtina, Kosovo, 2015 IEEE Paper.

[4] CVE Details: https://www.cvedetails.com/vulnerability-search.php

[5] Jonathan LeBlanc and Tim Messerschmidt, Identity and Data Security for Web Development, Published by OReilly Media, Inc. , 1005 Gravenstein Highway North, Sebastopol, CA 95472.

[6] John Paul Mueller ,Security for Web Developers, Published by OReilly Media, Inc., 1005 Gravenstein Highway North, Sebastopol, CA 95472.

[7] Miss R. V. Bhor, Prof. H. K. Khanuja, Analysis of web application security mechanism and attack detection using vulnerability injection technique, Department of Computer Engineering, MMCOE, Pune, India.

[8] Noor Ashitah Abu Othman Fakariah Hani Mohd Ali Mashyum Binti Mohd Noh, Secured Web Application Using Combination of Query Tokenization and Adaptive Method in Preventing SQL Injection Attack, Universiti Teknologi MARA Shah Alam, Selangor, Malaysia.

[9] OWASP: https://www.owasp.org/index.php/Top10-2013-Top10

[10] Ralph Adaimy, Wassim El-Hajj, Ghassen Ben Brahim, A Framework for Secure Information Flow Analysis in Web Applications, American University of Beirut, 2015 IEEE Paper.

[11] Shivangi Gu, Saru Dhir, Issues, Challenges and Estimation Process for Secure Web Application Development, Amity University Uttar Pradesh, 2016, 2016 IEEE Paper.

[12] William Bradley Glisson, L. Milton Glisson, Ray Welland, Secure Web Application Development and Global Regulation, University of Glasgow, 2007 IEEE Paper.

Dr.T.Subbulakshmi is currently working in VIT University, Chennai, India as professor. The author has 14+ years of experiencce of doing consultancy projects using FOSS. The author has introduced coursers using FOSS for UG and PG students. The author is involved in design of new operating systems based on Linux Kernel and member of SIG-OS, open source forums and mailing lists.